CONTENTS

WAKEHURST —
WILD PLANTS FROM WILD PLACES

Botanic garden and plantsman's paradise; wildflower refuge and ancient woodland; world-famous seed bank and conservation pioneer; Elizabethan mansion and Victorian country estate. Wakehurst is all of these things and more. And, it is deeply rooted in one of the best-preserved medieval landscapes in Europe.

For 50 years, this extraordinary estate of 465 acres has been cared for and developed by the Royal Botanic Gardens, Kew. This Guide tells Wakehurst's story and explores the very special collections of plants and places that you can discover.

Stuartia pseudocamellia

WELCOME TO WAKEHURST

Wakehurst is a garden for all interests and all seasons. There can be few locations in England with the rich variety of this Sussex estate, which is both cradled by an historic landscape and enriched by cutting-edge botanical and conservation science.

The estate is laid out to take maximum advantage of the dramatic topography (see A Changing Landscape, p58). From the highest point at the garden entrance, you are guided down through stately trees towards the grand Elizabethan Mansion, where the gardens open out in a great swirl of manicured lawns and borders. The gardens cascade east and west around ancient meadows, tumbling steeply down through Westwood Valley E 2 and Bethlehem Wood G 7 , slowing at the Pinetum E 4 and Bloomers Valley H 9 before dropping again to the sheltered, tranquil Westwood Lake B 4 . Managed gardens then give way to ancient and coppiced woodland, where rare dormice and English bluebells flourish, and kingfishers hunt on Ardingly Reservoir.

This broad portrait omits the detail, which includes the bright colours and reflections of the Iris Dell G2 , the soaring conifers – even now strikingly exotic against the backdrop of native woodland, the rhododendrons forming a wall of colour in the ravine, the cryptogams mottling the dramatic sandstone outcrops, the clusters of bee orchids near the Visitor Centre I 6 , and the beautifully tended Walled Gardens beside the Mansion G5 .

It was the enviable range of planting opportunities in Wakehurst's rich acid soils, shady ravines and sunny meadows, along with its benign climate, that attracted the Royal Botanic Gardens, Kew to adopt this estate as its sister garden 50 years ago.

The Elizabethan Mansion is set among beautiful lawns, rare trees and walled gardens.

You can see bee orchids (*Ophrys apifera*) near the Visitor Centre in late June and early July.

What makes Wakehurst a botanic garden?

A botanic garden has documented collections of living plants, which are used for scientific research, conservation, display and education. Wakehurst is renowned for its trees and shrubs collected from around the world, in particular from North America, the southern hemisphere and the mountain regions of Asia. It has National Collections of species of birch (*Betula*), southern beech (*Nothofagus*), *Skimmia* and St John's Wort (*Hypericum*). The plants, and details of where they were collected, are recorded in the garden's extensive database and this information is carried on their labels, which makes Wakehurst a living encyclopedia for plant lovers.

WAKEHURST TODAY

Wakehurst's estate of 465 acres is entirely managed by the Royal Botanic Gardens, Kew under lease from the National Trust. It is both a Grade II* listed heritage garden of great beauty and a beacon of botanical horticulture and nature conservation.

The estate includes the renowned Millennium Seed Bank (MSB) H6. Working with partners in more than 80 countries around the world, the MSB pushes technological boundaries to save wild plants by conserving their seed (see Saving the World's Seeds, p54).

Over the last 50 years, Kew's country garden has been managed to maximise both its native flora and its collections from around the world.

With a wetter climate, cleaner air and richer, more acidic soils than the Royal Botanic Gardens in Surrey, it provides ideal growing conditions for many trees and shrubs. Earlier plantings have been modified and extended to reflect geographical origins. A circuit of the estate takes you on an adventure through the temperate woodlands of the world, where you can explore the myriad forms and flowers of both the northern and southern hemispheres, and delight in the handsome shrubs of Himalayan hillsides, whose beauty disguises their toughness.

Through Kew's care, British wildflowers have flourished on the estate. Swathes of snowdrops, primroses, bluebells and orchids bloom in succession and rare meadowland species are increasing, encouraged by a return to traditional management practices (see p46). Age-old countryside crafts have reaped great rewards in the Loder Valley Nature Reserve A3 (see p50) – the first nature reserve to be established by a botanic garden. The mosaic of wetlands, meadows, coppiced and ancient woodland teems with birds and butterflies, wild plants, and animals such as badgers and dormice.

This sensitive approach has also conserved a habitat that has international significance thanks to the plant community it supports. Wakehurst is blessed with extensive outcrops of sandstone known locally as sandrock. Sinuous, richly coloured root systems of yew trees snake over many of the outcrops in the Francis Rose Reserve, and in their humid shade thrive many rare lichens, mosses and ferns (known collectively as cryptogams).

You'll find swathes of wildflowers in the woods at Wakehurst including our native bluebell.

Make sure you visit the
Millennium Seed Bank,
one of the most ambitious
conservation projects in
the world.

SEASONS AT WAKEHURST

SPRING

It can be hard to tell when winter ends and spring begins at Wakehurst, with the array of flowers gradually changing as weeks pass. From March a crescendo of naturalised daffodils and magnificent tree magnolias bring spring to the garden with a flourish. The magnolias in the Slips G3 and scattered through the woodlands of Westwood Valley E2 and Horsebridge Wood D7 stand out all the more because the surrounding trees are yet to grow their new leaves.

Spring bulbs are always a welcome sight and at Wakehurst the daffodils and narcissi planted in the lawns are joined by many other, often unusual, species. Close to the Mansion is the Monocotyledon Border, where bulbs and other plants classed as 'monocotyledons' are grown. Drifts of glory of the snow (*Chionodoxa*), snowflake (*Leucojum*), narcissi and tulip species bring this border to life.

RIGHT
Forsythia × *intermedia* 'Lynwood' sings out in the Spring Border.

LEFT
**Pheasant-eye daffodils
and many other spring bulbs
add colour to the lawns.**

**Wakehurst's magnolias
are one of the highlights
of spring in the Slips.**

By the Mansion Pond, you'll find a border planted to celebrate spring colour. A mix of shrubs and herbaceous plants crowd the edge of the pond, with *Forsythia*, *Corylopsis* and *Kerria* underplanted with violas, primulas and more unusual plants such as fawn lily (*Erythronium*) and shooting stars (*Dodecatheon*). The nearby Asian Heath Garden **H3** is also at its best at this time of year, with dwarf high-elevation species of rhododendron flowering in succession as the days warm.

The Water Gardens **F2** are now awakening from their winter slumber, and you should spot marsh marigolds and skunk cabbage dotting the beds with their vivid yellow flowers, with a backdrop of tree-like rhododendrons blooming above them. Rhododendrons are a major part of the spring garden displays, with Asian species throughout the Asian Heath Garden and larger species and varieties in the Water Gardens and through Westwood Valley.

As spring progresses, a succession of native wildflowers put on a show through the natural areas of the estate. From primroses in March, the display builds through spring to late April or early May, when thousands of English bluebells carpet the woodland floor through Westwood Valley and Horsebridge Wood. Such a sight is not to be missed and a walk through these woods lifts the spirits and offers the promise of warmer days to come.

SUMMER

From the beginning of June until early September, Wakehurst is a sea of vibrant colours and heady scents. The incredible Walled Garden G5 is a must-see throughout summer with sweet peas, roses and chocolate cosmos offering a delicious mix of scents, while the pastel colour scheme here is always restful. Next to it, the long West Mansion Border hugs the house itself and is planted with bolder colours in a rainbow of shades, while the formal Pleasaunce Walled Garden has North American prairie varieties to admire, including *Coreopsis*, *Helenium* and *Echinacea*.

Bordering the lawns by the Mansion, the Southern Hemisphere Garden F4 and Specimen Beds G4 are filled with interesting and colourful shrubs and herbaceous plants. Fascinating plants from the southern continents include beautiful white-flowered *Eucryphia* and brightly coloured red-hot pokers (*Kniphofia*). In the Specimen Beds, you'll find a wide variety of hydrangeas, penstemons, buddleias and other plants competing for attention, with some of the beds colour-themed to suggest ideas for your own garden.

Moving away from the Mansion, the Water Gardens F2 are a summer highlight. If you walk from the wildflower meadow of the Slips, past the Bog Garden and Pollination Garden to the magnificent Iris Dell, there are many glorious sights in this special part of the estate. The Water Gardens are planted with a mix of moisture-loving plants, so summer is the time for hostas, rodgersias, astilbes and myriad other interesting varieties. The Pollination Garden holds plants of value to bees and other pollinators, and is a source of great inspiration to gardeners. In the Iris Dell, you can delight in over 60 varieties of Japanese water iris, which fill the beds with flamboyant flowers from mid-June to mid-July.

Red-hot pokers add vibrant colour to the Southern Hemisphere Garden.

A stroll through the botanical collections in the woodlands reveals a wealth of exotic trees at their summer best, as well as many native wildflowers, which thrive beneath them. As you reach Bloomers Valley H9 or the area around the Millennium Seed Bank H6, specially created wildflower meadows are a magnificent sight throughout summer. Native plants are every bit as beautiful as their exotic counterparts, and conserving the local flora and fauna here is a major part of Wakehurst's work.

The Specimen Beds near the Mansion are full of plant ideas for your own garden.

AUTUMN

The end of summer can be almost imperceptible as autumn colour slowly begins to appear in August, and the flowering displays carry on until October. As the days shorten, the exotic and native trees both take on their autumn hues. Always first is the Japanese katsura (*Cercidiphyllum*) with its strawberry-scented pink and orange leaves marking the start of the wonderful season to come. A visit at this time of year means you can marvel at the ever-changing flamboyant displays of reds, yellows and oranges throughout the gardens and woodlands. As maples, hickories, rowans and the National Collection of birches (*Betula*) G7 adopt their full colours, Wakehurst's woodlands are well worth repeated visits through the misty autumn days.

Not that the flowers of the gardens are finished yet. The displays in the Walled Garden G5 and Specimen Beds G4 carry on through September and October, with Michaelmas daisies (*Aster*), meadow saffron (*Colchicum*), stonecrops (*Sedum*) and *Liriope* to name just a few of the plants in

The wonderful autumn hues of the Japanese katsura tree are accompanied by a sweet candyfloss aroma.

Autumn cyclamen carpet the woodland floor near the Visitor Centre.

full flower now. Tender perennials like dahlias, *Argyranthemum* and *Diascia* will continue to flower until the frosts come, so can still be in bloom in November in some years.

One of Wakehurst's biggest autumn attractions, and definitely not to be missed, is the mass planting of autumn-flowering cyclamen at the Visitor Centre 15 . Over 30,000 plants of *Cyclamen hederifolium*, all raised from seed in England, have been planted here in huge drifts under the trees. Depending on the season, they may start to flower as early as August, but reach their zenith in September and October, with sheets of pink or white shuttlecock flowers nodding over beautifully marbled leaves.

As autumn slips into early winter, the last of the leaves turn and fall. Amongst the last trees to put on a show are the maidenhair tree (*Ginkgo biloba*) and sweet gum (*Liquidambar*). Brilliant yellow leaves mark out the *Ginkgo*, while the sweet gum changes through deep maroon and orange to yellow before the leaves fall in November. At the same time, in beds near the Mansion, the startling yellow flowers of *Mahonia* appear amongst their primeval-looking spiky evergreen leaves, marking the start of the winter display.

WINTER

Winter may be a long and often dark season, but that does not mean it is without interest at Wakehurst, and in fact it is the ideal time to see some of the garden's best features. From panoramic views across the varied terrain, to the coloured barks and hardy flowers of the Winter Garden, there is always something to admire.

Near to the Mansion you'll find many seasonal jewels in the Winter Garden G4. With five informal beds planted specifically for this season, this is a real treasure trove of ideas. There is a mix of plants for flowers, coloured stems or leaves, and for architectural effect. For long-lasting colour, you cannot ask for more than the bright yellow flowers of *Mahonia* or the amazing sweet-scented purple blooms of *Daphne bholua*. Both these plants will bloom from November right through to March, making them of tremendous value where space is limited. There are large drifts of *Daphne* in Westwood Valley E2, rewarding those that journey that far with wafts of irresistible scent.

The Winter Garden is also home to winter flowering heathers and delightful witch hazels with their spidery, scented blossoms on bare branches in January. Alongside these you'll find dogwoods and willows with brightly coloured stems, evergreens with variegated foliage such as *Euonymus* and *Fatsia,* and berry-bearing shrubs including a National Collection of *Skimmia*.

For those intent on a longer walk, you should head to Bethlehem Wood G7 where Wakehurst's National Collection of birch is an incredible sight through winter. Low sunlight glows on their peeling bark and brightly coloured trunks, with shades from deepest chocolate brown to glistening silvery white to admire.

Glorious views across the wonderful Wealden landscape are at their best through winter. With most of the trees leafless, the true nature of the terrain is exposed, with deep valleys and high ridges to be enjoyed. From the Visitor Centre, Bethlehem Wood and Coates' Wood I9, as well as across Westwood Valley and within the Loder Valley Nature Reserve A3, panoramic vistas are opened up for you to enjoy the landscape of Wakehurst and the surrounding countryside.

The sweet scent of *Daphne bholua* drifts up Westwood Valley in the winter months.

The bright yellow flowers of *Mahonia* bring much needed colour at this time of year.

FORMAL GARDENS

Formal gardens surround the Mansion at Wakehurst, although 'formal' is a relative term and, except for the clipped box and yew hedges, none of the garden features are truly formal. As you move away from the house the degree of informality increases, helping you acclimatise to the relaxed atmosphere that the gardeners here strive to create, and the peace and tranquillity of the woodlands and nature conservation areas.

The most intense horticultural and ornamental garden plantings are all within easy reach of the Mansion and the Stables restaurant. They are accessible to all and ideal for a gentle walk at any time of year.

THE WALLED GARDENS & MANSION BORDERS

The Walled Gardens G5 are amongst Wakehurst's finest features. You can take a stroll through two walled areas here and each has a different character, with the Victorian formal yew hedges of the Pleasaunce Garden contrasting with the informal cottage garden style of the Henry Price Memorial Garden.

Designed and constructed in the 1970s, pastel shades dominate the cottage garden planting of the Henry Price Memorial Garden, with roses and other shrubs intermingled with herbaceous perennials, tender plants and annuals. Colour and scent fill the air in summer, but you'll also find tulips extending the season into spring, and many of the tender plants then continuing to flower late into autumn.

From early May, when the tulips and alliums are in full flower, there is always something to delight the senses in the Walled Gardens. Roses, trained into columns or over low domes, are covered in scented pink flowers in June and July, with varieties of sweet pea and chocolate cosmos joining them as they come into flower for the rest of the summer. Traditional herbaceous perennials, such as phlox and Michaelmas daisies, are mixed with tender plants that are propagated in the nursery and planted out for summer. These include *Argyranthemum*, penstemons and South African *Diascia*, as well as more unusual plants like *Corethrogyne* from California.

Outside the Walled Gardens, a long border runs against the wall of the Mansion. This is the West Mansion Border and is planted in a rainbow of summer colour in true Gertrude Jekyll style. Using a mix of herbaceous perennials and tender plants, the border runs from oranges through reds, pinks, mauves and blues to white at the Walled Gardens gate. This really is a riot of colour and a glorious place to soak up the afternoon summer sun.

At right angles to the Mansion runs another border using the south-facing wall of the Walled Gardens as a backdrop. Here is Wakehurst's only taxonomic display. All the plants in this border are monocotyledons. These are plants with a single seed leaf and are the subject of much of Kew's scientific research. Here you'll find bulbs, grasses, orchids and palms. In keeping with Wakehurst's geographic planting elsewhere, the border is split into sections for plants from the northern and southern hemispheres.

Michaelmas daisies put on a vibrant show in the Walled Garden.

The Henry Price Memorial Garden is filled with colour and scent throughout summer.

THE SOUTHERN HEMISPHERE GARDEN

Sprawling out in front of the Mansion are carefully tended lawns over two levels. At the far side of the lower lawn you'll find a series of island borders that form one of Wakehurst's most intriguing garden features.

In the early 1900s Gerald Loder (Wakehurst's owner at the time) added wild-collected plants from the southern hemisphere to a mixture of trees, heathers and dwarf rhododendrons here. This inspired an expansion on the theme and has resulted in the Southern Hemisphere Garden F 4 .

In keeping with the planting scheme (grouping plants by geographic origin rather than by their relationships), the Southern Hemisphere Garden features plants from New Zealand, Australia, South Africa and South America in specific beds. Apart from the many interesting plants found on these continents including eucalyptus, hebe and unusual conifers, there are several iconic plant groups that are endemic to (or only found in) these lands. The protea family is an interesting example, and here you'll find several representatives including *Telopea*, *Hakea* and *Banksia* from Australia, as well as *Embothrium* from South America and *Protea* from South Africa. Look out too for the small genus of white flowered shrubs known as *Eucryphia*, which are only found in South America and Australia, while the more familiar *Fuchsia* hails from South America and New Zealand.

You'll find several species of *Eucryphia* from South America and Australia in the garden.

The flaming red flowers of *Embothrium* earned it the common name of Chilean fire bush.

The ancestors of these plants were once found across a single supercontinent – Gondwana. The presence of closely related plants in the isolated parts of this former landmass, but nowhere else on Earth, provides a living demonstration of the passage of time and movement of continents. As a botanic garden, this is something that can be easily shown in the collections for all visitors to marvel at.

Most of the plants that have been introduced to these beds are shrubs. However, you'll also spot some herbaceous plants including red-hot pokers (*Kniphofia*) and *Agapanthus* in the South African bed, and the extraordinary *Fuchsia procumbens* – a ground-cover plant from New Zealand. On some of these unusual species, the flowers are quite small and can be a challenge to find, but they are well worth seeking out. Indeed, there are many fascinating plants to search for in these beds with flowers appearing from early spring through to late autumn.

THE WATER GARDENS

Walking along the path away from the front of the Mansion brings you to an ornamental sundial G4 commemorating Gerald Loder. From here, a shallow valley called the Slips G3 runs to the head of Westwood Valley E2.

The beds below the sundial face due south and are planted with rock roses (*Cistus*), Jerusalem sage (*Phlomis*) and other plants seen around the Mediterranean Sea. Water cascades along a stream between large magnolias, which are a magnificent sight in early spring. The banks on either side are carefully managed for spring bulbs and native wildflowers, forming a corridor of colour for several months into summer.

Below the stream is the first of several ponds and water garden features. Black Pond is home to several large koi carp. It is surrounded by dogwoods grown for their coloured winter stems that reflect beautifully in the water. The stream continues in a cascade from this pond through beds planted with moisture-loving plants such as candelabra primulas, day lilies (*Hemerocallis*) and the spectacular giant Himalayan lily (*Cardiocrinum giganteum*).

The water then falls into another pond at the centre of the Iris Dell G2. This feature, developed in the 1990s, uses over 60 varieties of Japanese water iris (*Iris ensata*) as a mass planting that bursts into vibrant colour in late June and early July. These gorgeous plants form a carpet of blues, mauves and purples, but they only last a few brief weeks. Don't worry if you miss them though, you can still see the beautiful yellow flowers of *Narcissus cyclamineus* in March, evergreen azaleas in April and autumn colour on the coral bark maples (*Acer palmatum* 'Sango-kaku') in September.

Don't miss the spectacular magnolias which line the Slips.

Below the Iris Dell you'll find the main water garden, packed with a mixture of moisture-loving perennials including *Rodgersia*, *Astilbe*, primulas, irises and day lilies, as well as less common plants such as the Japanese *Kirengeshoma palmata*.

The Water Gardens are also home to the Bog Garden and Pollination Garden. Found opposite Black Pond, these two features combine ornamental planting with educational messages. The Bog Garden has been set out with a boardwalk to help school groups when pond dipping. It's also home to a number of British native plants showing the value of our own flora in gardens. The Pollination Garden combines a profusion of bee-friendly plants around working beehives and displays of beekeeping equipment. Here you'll find everything you need to know about pollen, pollination and pollinators.

Sixty varieties of Japanese water iris burst into bloom in the Iris Dell each June.

THE ASIAN HEATH GARDEN

The Tony Schilling Asian Heath Garden H3, to give this area its full title, was created in the aftermath of the 1987 'Great Storm'. The large rhododendrons that were growing here before that date were swept aside by the hurricane-force winds, so, when it came to replanting, the horticultural staff looked for more resilient species.

To help create a new garden, they looked to the high elevation rhododendrons that will tolerate sun and more exposed conditions, and devised a planting plan that mimicked the plants of the high Himalayan mountains.

The beds are set out with plants from China and the Himalayan chain, as well as those from Japan, Korea and Taiwan. Most of the plants growing here have been raised from seed or cuttings of wild plants, so they genuinely represent the flora of these mountains, particularly those growing at or above the tree line. This means there are few trees here, but you'll find large groups of shrubs, including a variety of rhododendrons, junipers, potentillas and cotoneasters.

The best time to visit is in spring when all the rhododendrons are bursting into bloom, but there are also interesting specimens for other seasons, including winter-flowering *Lonicera setifera*, summer lilies and species hydrangeas, and *Roscoea* and *Cautleya* in flower in autumn.

The rhododendrons in the Asian Heath Garden put on a spectacular display.

Next to the Asian Heath Garden, a narrow border runs alongside the Mansion Pond. This is the Spring Border and, as the name suggests, it is planted with shrubs, perennials and bulbs that are at their peak through the spring months. From March onwards, myriad plants mark this border out as a must-see part of the estate; from crocuses and snowdrops naturalised in the grass to swathes of daffodils and the beautiful fawn lily (*Erythronium*), a succession of flowers can be seen here until late May.

At the north end of the Spring Border is a small rocky outcrop. Known as the Rock Terraces, this feature was established by Lady Price (see The Making of the Gardens, p59). The most impressive plants here are the several venerable Japanese maples dating from that time, with their weeping branches and intricate dissected leaves. Other small trees and alpine plants are grouped together here in this small but beautiful part of the gardens.

Trollius 'Orange Princess' is just one of the delights of the Spring Border.

The Great Storm

Overnight on 16 October 1987, a deep depression swept from the Bay of Biscay over southern England. Hurricane-force winds destroyed 15 million trees across the country, with around 20,000 trees lost at Wakehurst. This devastation took many years to recover from and Wakehurst lost some old and irreplaceable specimens.

At the time this seemed like a disaster, but looking back now it is obvious it was not all bad news. The loss of old trees created space to replant. Collections could be refined using opportunities to gather wild plants from wild places around the world to improve this botanic garden.

THE WINTER GARDEN

Wakehurst is open all year round and there are many highlights to enjoy in spring and summer. However, you'll also find a surprising amount to enjoy in winter too, including the dedicated Winter Garden G4.

Here, in five separate beds close to the Mansion, you can explore interesting combinations of plants that flower through winter as well as those that have attractive coloured stems or bark, or have interesting architectural shapes.

Since this garden was created in the 1980s the planting has been gradually refined, reducing the number of different varieties used, but expanding the size of each group, to create a bolder and more attractive layout. Large informal groups and ribbons of plants intertwine colours and textures across each bed. New colour combinations and themes now run through the most recently replanted beds.

Winter box fills the air with amazing scent.

Many key plants thrive here including the beautifully scented witch hazels (*Hamamelis*), winter box (*Sarcococca*) and winter honeysuckles (*Lonicera*), as well as heathers and winter bulbs such as snowdrops (*Galanthus*). Alongside the flowers, you'll not fail to notice the dramatic coloured stems of dogwoods (*Cornus*), willows (*Salix*) and the more unusual black stems of *Hydrangea macrophylla* 'Nigra'. Among the beds there are also feature trees, including birches, cypress and golden pine (*Pinus sylvestris* 'Aurea').

The coloured stems of dogwood and willows add drama to the Winter Garden.

From November, when the spikey-leaved mahonias develop their dazzling yellow flowers, through to March when camellias and rhododendrons finish the show, there is always something new to see. Wakehurst was amongst the first places in Britain to grow the Himalayan *Daphne bholua*, with its mauve highly-scented flowers. Once rare in our gardens, it now more widely available, but requires acidic soil to thrive. By the time the Winter Garden fades in late March, many other areas of the garden are coming into their own, so this feature carries you into the brighter days of spring.

This is one of the country's oldest and best-loved winter gardens and plays to one of Wakehurst's unique strengths. You'll discover interesting combinations of plants common in gardens with wild species and rarities in this attractive and also scientifically valuable garden. This type of planting is important for Wakehurst's collections and clearly sets the garden apart from other country estates.

NATIONAL PLANT COLLECTIONS

Wakehurst is home to four National Plant Collections under the Plant Heritage scheme. This aims to conserve plants in gardens across the country, with public gardens and enthusiastic amateurs alike each taking plants under their wing.

Wakehurst has two National Collections of shrubs: shrubby *Hypericum* (technically, species in sections *Androsaemum* and *Ascyreia* of the genus) and *Skimmia* in the upper gardens. There are also collections of two genera of trees: birches (*Betula*) and southern beech (*Nothofagus*) in the woods. If you don't have time to visit all of these collections

you can see examples of all four in a single bed next to the Stables restaurant H4.

The golden yellow flowers of *Hypericum* (St. John's wort) form a large part of the summer display in the Specimen Beds and are highly attractive to bees, including those from the garden's own hives. These plants have hidden values too, as researchers

You can find the National Collection of *Hypericum* in the Specimen Beds.

using this collection have discovered. One species here holds a chemical that combats infections caused by methicillin-resistant *Staphylococcus aureus* (MRSA), while in many European countries, solitary bees take wax from the flowers to line their underground egg chambers, protecting their eggs from summer rain.

Skimmia is related to oranges and lemons, but rather than producing large juicy fruits these small shrubs put on a show of small bright red or white berries over winter. They also bear heads of fragrant flowers in spring and with their glossy leaves they are a valuable year-round addition to any garden. *Skimmia* has separate male and female plants, which means you need to plant both to get a display of berries, so this National Collection is clustered together in beds near to the Mansion.

Birches are trees of the northern hemisphere. They are pioneers of new habitats and are highly variable. Wakehurst's collection of *Betula* species in Bethlehem Wood G 7 shows just how many bark colours and different leaf shapes they have, and has been used for reference in the latest monographic study of the genus, published by Kew in 2013. Southern beech (*Nothofagus*) by contrast is an exclusively southern hemisphere genus, with species native to South America, Australia and New Zealand. These graceful trees can be evergreen or deciduous, depending on species. You'll find individual specimens in the Southern Hemisphere Garden, but most of the collection is in Coates' Wood I 9 .

Discover the National Collection of birch in Bethlehem Wood.

The fragrant flowers of *Skimmia* turn into masses of bright red berries in autumn.

NATURAL GARDENS AND WOODS

One of the main attractions of the woodlands and conservation areas at Wakehurst is their peace and tranquillity. These more informal areas are places to relax, unwind and enjoy nature, and Wakehurst's horticulturists work hard to ensure they will always remain so.

 The trees here are planted out according to the country they are native to, as well as by climate and the other plants they naturally grow with. This careful arrangement means you can go on a unique journey through the temperate woodlands of the world without having to travel very far.

 Many of the trees you'll see here are the result of Kew's horticulture and science teams coming together to collect seed from wild plants around the world. The woods are also managed to encourage wildflowers, native plants and animals, so you can enjoy a beautiful woodland experience at any time of year.

HORSEBRIDGE WOOD

Wakehurst's botanical collections are set out according to the country or continent that the plants are native to, to complement the taxonomic collections at Kew.

Rather than grouping related plants together, plants at Wakehurst are placed according to where they grow naturally. This is called a 'phytogeographic' scheme ('phyto' simply means 'plants'). Such planting helps concentrate conservation work on areas with the greatest diversity of unique plants.

Horsebridge Wood D 7 is an excellent example of this system in practise and is home to trees from North America. This is a huge continent, so habitats vary significantly and the planting reflects this. Trees are set out by regions, so as you walk from Westwood Lake B 4 toward the Millennium Seed

Bank H 6 you pass through the trees from the east coast of the USA and Canada, then from California to the Pacific west coast and Rocky Mountains.

Each region has a distinct flora, so the east coast, or Appalachian region is dominated by deciduous trees and their spectacular 'fall' colours. Trees such as sweet gum (*Liquidambar styraciflua*) and hickory (*Carya tomentosa*) bring this woodland to a climax of colour in October each year. They are also the subject of Native American stories, and essential for the survival of many native animals, and this is celebrated in the 'Talking Totems' natural play spaces in Horsebridge Wood, just two of a series of these spaces spread around the estate for families to enjoy.

Further into Horsebridge Wood the vegetation changes to coniferous woodland and the fascinating plants of the Californian region, which enjoys a Mediterranean climate. You then reach the Pacific coastal conifer forests of the Vancouver region, dominated by stately redwoods, hemlocks and Douglas fir (*Pseudotsuga menziesii*). Although these are not exact recreations of the forests of these regions, especially as the woodland floor is covered with UK native wildflowers, you can still get a great insight into the changing forests of the temperate regions of the world.

Wakehurst's wildflowers are precious. Horsebridge Wood is justly renowned for the spectacular display of thousands of bluebells flowering here each April and May (see right). This magnificent sight is well worth the walk in spring, but you can also look out for other flowers here later in the year, such as goldenrod (*Solidago virgaurea*) and devil's bit scabious (*Succisa pratensis*).

The sweet gum puts on a riot of colour in late autumn.

BETHLEHEM WOOD

This woodland area G7, close to the Millennium Seed Bank, is home to a National Collection of birches (*Betula*). These stunning trees are pioneers from the northern hemisphere, where they colonise clearings and newly opened habitats as well as regions where larger trees cannot survive, such as the arctic tundra.

Some tundra species are almost impossible to cultivate at Wakehurst, where warm winters and late frosts can kill their young foliage. However, a wide range of birch species grow well here and this collection is one of Britain's finest. These are beautiful trees at any time of year – their coloured bark glows in the winter sunlight, while their delicate leaves provide dappled shade through the summer months. As summer ends, many of the American species adopt rich yellow autumn colours, creating a wonderful woodland spectacle.

You can see the wide variety of species at close quarters throughout the year by following the birch trail, mowed into the grass through the collection. The colour of the bark can vary dramatically within each species, so differences in the trees' flowers and leaves are better for botanists to use to distinguish between variants. The trees are set out geographically, so each section of Bethlehem Wood has birches from a different part of the northern hemisphere.

Families can enjoy several natural play spaces at Wakehurst including 'Trunk Hopping' here in Bethlehem Wood.

Like all of Wakehurst's woodlands, Bethlehem Wood is carefully managed for native wildflowers. The grass and flowers are cut in late summer or autumn after they have produced seed, and the clippings are removed, so that populations are maintained and improved. This woodland has the highest diversity of wildflowers of any part of Wakehurst, with over 125 different species to be seen. Primroses, lady's smock (*Cardamine pratensis*) and bluebells flower in succession through spring, beautifully complementing the trees.

You will find one of the natural play spaces here too. Nestled amongst the trees is 'Trunk Hopping', which takes the theme of partners from the nearby Millennium Seed Bank in a three dimensional trail for all ages, leading to a carved treasure chest.

COATES' WOOD

Coates' Wood I 9 is named after Gerald Loder's head gardener, Alfred Coates. It was amongst the worst affected parts of Wakehurst following the 1987 storm, with many fine specimens lost to the wind. After the clear up a goodly-sized area was available for new collections, and, following Wakehurst's geographic planting scheme, species of southern hemisphere trees were placed here.

Walking from Bethlehem Wood, you enter the woodlands of Australia and New Zealand, dominated by *Eucalyptus*, a grove of Wollemi pine (*Wollemia nobilis*) and southern beech (*Nothofagus*). A shelter-belt established along the edge of this woodland after the storm protects these collections from any future high winds, while another of the natural play spaces, 'Root Route', is part of a small picnic area sheltered amongst the trees here.

There are some glorious views from this woodland. From the Australian section you can look out across Bloomers Valley H 9 to admire the wildflower meadow with its patchwork of summer colours. From the South American section there are longer views over Bloomers Valley to Horsebridge Wood beyond.

As you move around the horseshoe shape of Coates' Wood, you pass into the trees of South America, with groups of fascinating plants including *Fitzroya cupressoides*, a conifer named for Captain Fitzroy, commander of the HMS *Beagle* on Charles Darwin's famous voyage around South America. Known as alerce, it is South America's largest tree, but is endangered by logging activities. It is now illegal to trade in its wood. A shipment impounded by HM Customs and Excise was donated to Kew to build the Field Study Centre B 3 for schools at the bottom of Westwood Valley. Other plants to look out for here include *Eucryphia*, *Drimys* and *Lomatia*. All of these shrubs have attractive flowers, making Coates' Wood a beautiful place to walk and admire the collections.

Wakehurst's National Collection of southern beech (*Nothofagus*) is also here. These intriguing trees are distant cousins of our native beech (*Fagus sylvatica*), but are exclusively southern hemisphere trees. Large groups of each species have been planted to develop a woodland dominated by these trees, as happens in their native homelands.

Lomatia hirsuta is among the rarities you can see in Coates' Wood.

THE PINETUM

A pinetum is simply a collection of conifers. Wakehurst's Pinetum E4 is set out where Gerald Loder planted his second conifer collection in the 1920s, but, in common with many parts of the garden, it has evolved under Kew's stewardship. The conifers you'll find here now are planted out in a geographic system, with trees from the same regions grouped together.

The Pinetum was devastated by the 1987 storm, and lost around 80% of its mature trees. In many ways this allowed a certain freedom to redevelop the collection, as old specimens would never have been removed through choice. Globally, there is a larger land area under conifer forest than any other vegetation type, and yet these plants are often overlooked and outshone by their flowering relations. Huge tracts of the northern hemisphere in Asia and North America are home to conifers, while several fascinating species are only found on the southern continents.

Interesting specimens to be seen in the Pinetum include the Wollemi pine (*Wollemia nobilis*) and *Podocarpus* from Australia, and *Taiwania* and Japanese cedar (*Cryptomeria*) from Taiwan and Korea. One large specimen of *Cryptomeria japonica* has a carving of a sika deer stag's head set in its trunk. This traditional style of carving (called Tachigi-bori) has been practised in Japan for centuries. The trunk of a living tree is carved to a design, but the bark is allowed to slowly enclose the carving to show the passage of time. This carving was expertly completed by artist Masa Suzuki in 2013 as part of a Japanese-British collaboration, into a wound left by the 1987 storm.

The Korean fir, *Abies koreana*, is a popular ornamental conifer of cool climates.

During the Second World War, Wakehurst was home to the 1st Canadian Corps. As part of their interaction with resistance forces in France, an underground communication station was built in the Pinetum. It remains there to this day, although the passing years have taken their toll and it is no longer viewable. Its position is marked and you can read all about this intriguing part of Wakehurst's history here.

ABOVE
This traditional Japanese carving in the dead wood of a living tree explores people's connection to nature.

RIGHT
The tamarack, *Larix laricina*, is native to Canada.

THE HIMALAYAN GLADE

Leaving the Pinetum you reach the edge of Westwood Valley. Turn to the right and you soon come across the dramatic Himalayan Glade D3 . Although this is built around two natural outcrops of sandstone, much work has been done here to develop an informal garden feature.

In the 1980s, Wakehurst's curator, Tony Schilling, wanted to recreate a small section of the Himalayan mountains – his first botanical love. By using Wakehurst's dramatic landscape and by adding features, a small sheer-sided valley was turned into a simulation of a high mountain pass, with *Berberis* and other dwarf shrubs surrounded by the rock faces. As the valley tumbles down towards the stream, larger shrubs such as rhododendrons appear, alongside perennials from these mountains including Himalayan knotweed (*Aconogonum*).

Maintaining a Himalayan mountainside in Sussex is an odd occupation for the garden staff. Many of the shrubs here would be grazed by yaks in their homeland – something that is lacking in these woodlands! So, periodically the *Berberis* and other shrubs are cut almost to the ground to allow these compact dwarf plants to grow back. The grass is also cut throughout summer to mimic grazing animals.

The Himalayan Glade is also popular with native birds and for many years visitors have fed the birds here. The birds have learnt to appreciate their hand-outs and many species can be seen around the post and rail fences at the top of each sandstone outcrop.

Berberis wilsoniae is densely planted in the centre of the Himalayan Glade.

Take a stroll through the Himalayan mountains to see many unique plants.

On the opposite side of Westwood Valley, strictly outside the Himalayan Glade itself, a stone seat is built into the hillside. This is part of the Himalayan re-creation, built by the garden staff and designed in the same way as the stone seats built for Sherpas in the mountains, with two separate levels: one to sit on, and one for their load to rest upon to take the weight off the Sherpa's back. This version has planted specimens of *Juniperus recurva* var. *coxii* next to it, as these seats do in the Himalaya, and around it are drifts of *Daphne bholua*, filling the air with sweet scent throughout winter.

WESTWOOD VALLEY

The steep valley that links the Water Gardens F2 with Westwood Lake B4 is known as Westwood Valley E2. This is one of the most dramatic landscape features at Wakehurst and is home to the East Asian botanical tree collections.

Following the geographic planting system, trees and shrubs from regions of China, eastern Russia, north India, Burma, Japan, Korea and Taiwan are grouped together here in a native oak woodland. The planted additions enhance the natural landscape, and throughout spring the magnolias and rhododendrons scatter the Valley with flowers.

The majority of *Rhododendron* species originate in Asia, so it is here that you'll find many of the large tree-like plants of this genus, sheltered by the deep contours of the land. They share the Valley with many interesting trees, raised from seed in Wakehurst's own nursery. Collaboration between Wakehurst's horticulturists, the Millennium Seed Bank, the Forestry Commission and others has allowed many species to be grown from wild-collected seed and added to the collections here. From China these include the Chinese tulip tree (*Liriodendron chinese*) and unusual winter-flowering *Illicium simondsii*. From Japan there are maples (*Acer*) and beautiful flowering dogwoods (*Cornus*),

and from Taiwan there is a unique conifer, related to redwoods, appropriately called *Taiwania formosa*.

In the deepest part of the Valley, below the Himalayan Glade, is Wakehurst's tallest tree. A Douglas fir (*Pseudotsuga menziesii*) grows close to the stream and towers over many of the other specimens (see image, right). It reaches a height of over 43 m (140 ft), but because of the terrain here the sheer height of this tree is not obvious. It is best seen from the south side of Westwood Valley, from the stone seat built into the hillside.

Paths running along the top of the Valley give you a good view of all the collections. You will come across the garden's largest natural play space on one of these paths – 'Unexpected Endings'. This brick labyrinth represents the convoluted patterns on the base of a pine cone, combining intriguing science with relaxation – a theme throughout Wakehurst. Some of the garden's best views are to be had along these paths in winter, with the leafless trees allowing glimpses across Westwood Lake into the surrounding countryside.

Follow the 500-metre-long path round the labyrinth of 'Unexpected Endings'.

WESTWOOD LAKE

Westwood Lake B4 marks the lowest point of the garden, although the Loder Valley Nature Reserve A3 stretches on towards Ardingly below this. This substantial body of water is actually artificial and was formed by damming Ardingly Brook.

The Lake is now not only a beautiful part of Wakehurst, but also the source of irrigation water for the garden and nursery. The small oak building at the lake-edge looks like a boathouse, but actually holds pumps, sending water to the upper gardens.

Westwood Lake and its surroundings are amongst the most beautiful and valuable environments in the gardens. There is a tranquillity here that makes this part of Wakehurst a wonderful place to visit. The areas north and south of the Lake itself are wetland habitats, full of interesting plants and animals. Frogs make their home here amongst the marsh marigolds and bur-reeds, while kingfishers can be seen occasionally from the raised boardwalk, darting along the stream as it enters the Lake. In the Wetland Conservation Area A3 to the south of the Lake you'll see coppiced hazel and stooled willows, which are cut and used for plant supports in the gardens. There is another raised boardwalk here, to get closer to the water and for pond dipping. This is used through much of the year by school groups based at the nearby Field Study Centre B3.

The stream flowing into Westwood Lake carries a great deal of suspended sand and silt, which is why the water of the Lake is rarely clear. This silt means that the Lake has required dredging and de-silting three times in the last 60 years. The most recent dredging, carried out in 2011, also saw the introduction of a silt trap toward the north end of the Lake. This slows the flow of water and much of the suspended silt falls to the lake-bed. In time this will mean the smaller upper section of the Lake requires more frequent dredging, but the main water body should remain clear for much longer.

BLOOMERS VALLEY & THE MEADOWS

Between Horsebridge Wood D7 and Bethlehem Wood G7 there is an open area, deliberately not planted with trees. This open grassland is Bloomers Valley H9, which affords magnificent views between Horsebridge and Coates' Woods, opening vistas that would otherwise be lost if this area was planted.

This area was ploughed for vegetable production during the Second World War, and then retained as mown grass for many years after that. Over the last ten years, the area has become a wildflower meadow. This was not as simple as stopping mowing, as the right mixture of species is needed. With careful management the wildflowers already present on site have seeded and spread through the grass. Selected wildflower species native to the Weald of Kent and Sussex have been raised in the nursery as plug plants and planted out. Amongst the mosaic of wildflowers you will also now find less common species such as dyers greenweed (*Genista tinctoria*) and saw-wort (*Serratula tinctoria*).

These are not the only meadows that have been created on the estate. The ground surrounding the Millennium Seed Bank H6 was landscaped into gentle contours when the building was constructed. It was deliberately planned as a patchwork of small wooded areas and wildflower meadow, to highlight the conservation work carried out within the UK by the Millennium Seed Bank Partnership and to blend the site with the rest of Wakehurst. Seed collected in the Loder Valley Nature Reserve A3 was introduced after pernicious weeds like docks had been removed.

Then plug plants were added to increase the diversity of the meadow. So you'll now find a wide variety of wildflowers there, from butter-yellow cowslips (*Primula veris*) in spring to vibrant purple devil's bit scabious (*Succisa pratensis*) in late summer.

Around the margins of Bloomer's Valley you'll find woodland areas representing a wide range of regions that are home to many fascinating species. Here you can hunt for unusual specimens from the 'circumboreal forests' that encircle the Arctic, those of the Mediterranean, and those of a region between the eastern Mediterranean and the Caucasus called the Irano-Turanian region.

An intricate mosaic of species is found in the meadows.

Keeping the garden buzzing

The garden team manage a multitude of tasks. Where else could you find a gardener who is also a trained beekeeper, managing our hives for the health of the bees and to produce our own honey? Beekeeping is a skilled job, as hives are subject to many pests and diseases, while managing the bees' habit of swarming needs careful observation and timing.

Having a team member qualified and skilled in both gardening and beekeeping is a huge advantage for Wakehurst. They are able to grow plants that are good for bees and keep those bees healthy to pollinate the plants.

Traditional hay harvesting
helps to increase the biodiversity
of the meadows.

CONSERVING WILD PLANTS

Although visitors appreciate the beautiful ornamental gardens and planted woodlands, Wakehurst is much more than either of these. Conservation is at the heart of the garden, whether it is in maintaining rare plants from temperate climates, banking seeds of threatened plants globally, or preserving the local environment.

Much of the work of the horticultural teams is centred on habitat management on the estate. Wakehurst has its own flock of Southdown sheep, introduced to help improve the wildflower meadows. Sheep do not cut grass like a mower, they are selective in what they eat and trample the ground as they move. This opens spaces for seed to germinate and ultimately enriches the meadow. This requires someone to watch over the flocks, so one of the conservation team also acts as shepherd. There are also hives of bees, kept to help pollinate the flowers across Wakehurst – both introduced ornamentals and wildflowers alike. Beekeeping is a complicated pastime, and a dedicated member of the garden team is also the beekeeper, tending the hives and keeping them free of pests and diseases.

Improving the native woodland is also important. Through winter, blocks of hazel are coppiced, producing pea sticks and poles for supporting herbaceous plants in the gardens. Coppicing is

Southdown sheep help
to manage the meadows.

good for wildflowers and animals as well, so this is an important way to maintain and enhance the environment. Coppiced wood is also used for charcoal, made using a kiln, this removes surplus wood and creates a useful product – Bar-B-Kew charcoal!

Through winter the staff also lay hedges. Hedge-laying is a difficult skill to master, but amongst the conservation team there is an experienced hedge-layer. This natural way of managing a hedge lowers its height, but also thickens it, bringing older shoots down at an angle to re-grow. This makes hedges stock-proof and also provides more places for birds to nest-build.

There is much more to Wakehurst than meets the eye. Staff monitor the wildflowers, butterflies, bats and birds across the estate. 'Horticulturist' is perhaps too narrow a title for most of the staff here, as they do much more varied work than in most gardens.

Traditional management such as hedge-laying is in evidence throughout the gardens.

A sustainable product

Managing woodlands is a traditional country craft, so something an estate like Wakehurst must do. Much of the woodland in the Loder Valley Nature Reserve is coppiced, cut on a long-term cycle to maintain a sustainable supply of timber. Conservation to improve these woodlands is the work of the reserve warden.

Much of the timber is not usable for woodwork so another use is needed. Using Wakehurst's own kiln, the warden takes surplus wood through a very messy process to recycle it into charcoal, which you can buy at the Visitor Centre. This work produces a valuable product and improves the estate's habitats at the same time.

THE LODER VALLEY

Wakehurst was one of the first botanic gardens in the world to open an adjoining nature reserve for native flora and fauna, starting a trend that has become widespread as gardens have come to value their own plants as well as those from abroad.

The Loder Valley Nature Reserve A3 was opened in 1980, on land passed to Wakehurst by the local water authority after the construction of the Ardingly Reservoir. The Reserve consists of a matrix of habitats, with wooded steep-sided valleys (ghylls) and open meadow areas, as well as the damp margins of the reservoir, all encompassed within its 16 hectares (40 acres). Wildlife abounds and you'll find bird hides at strategic places to allow you to view herons, kingfishers and water birds on the reservoir. There is also a badger hide placed by a large sett, where you can attend pre-arranged badger-watching evenings. To book a place on these thrilling sessions contact the Visitor Services team at Wakehurst (see details at the back of this guide). Wildflowers are encouraged and managed here too of course and the Reserve offers an enormous variety of native plants.

Much of the conservation activity is centred in the Loder Valley, managing habitats for the benefit of the wildlife here. Regular coppicing, hedge-laying, ride-widening and meadow-mowing all contribute to the biodiversity of the reserve. Recording the effects of any activity on the flora and fauna is an important part of the management of this site, so flowering plants and wildlife are monitored throughout the year.

Entry to the Reserve is via a locked gate next to the Field Study Centre B 3 , so you will need to obtain the key code at the Visitor Centre I 6 before setting off. It is a long walk through the gardens before you reach the Reserve, as well as a long walk around its lovely natural environments. It is therefore wise to be prepared for changing weather and for the long walk back.

There are around 30 species of butterfly in the Reserve, as well as 100 species of bird, 300 species of plants and many mammals such as badgers and dormice.

Caring for the land

Wakehurst is a country estate. This means the land is managed as well as the gardens and woodlands. So here you'll find an arborist who is also a shepherd. The day is spent monitoring and pruning our mature trees to ensure they are safe and in good condition, but mornings and evenings are for checking the flock of Southdown sheep.

The sheep are very good at selectively cropping the meadows to improve biodiversity, but they are prone to every ailment imaginable. So a great deal of skill and precision is needed to care for them, something Wakehurst's arborist-shepherd has to have in abundance.

CONSERVATION

Wakehurst has long been managed to encourage the native plants and animals that we share the land with, particularly since the opening of the Loder Valley Nature Reserve A3 and later the Francis Rose Reserve D3 for lower plants (cryptogams). As a result you'll find a huge diversity of plants and animals here.

From open meadows through managed coppice woodland to dense oak woodland and waterside marsh, there is a wide range of habitats at Wakehurst. Steep-sided valleys with small streams at the base (ghylls) provide shaded and sunlit slopes, suited to different species. This range of situations is one of the reasons Wakehurst developed as a fine garden and arboretum, but also why there is huge natural biodiversity. Both the exotic and native species here are truly 'wild plants from wild places'.

Management of these habitats is key to their success. The coppice woodlands are cut in rotation to create sunlit patches amongst the trees, ideal for primroses, early purple orchids (*Orchis mascula*) and dormice alike. Dormice move from place to place in the canopy of coppiced hazel, so maintaining the hazel helps the dormice. In the same way, building an artificial kingfisher nesting bank has helped these jewel-like birds to thrive, while improving the health of fish stocks in the streams.

Removing alien invaders is also important. Himalayan balsam (*Impatiens glandulifera*) used to choke many of the streams, so removing these annual plants before they shed their seed has helped the native waterside plants. Ragwort and docks are removed from the meadows to avoid them spreading uncontrollably.

Wakehurst is a wonderful place to visit for anyone with an interest in native wildflowers. Throughout the year there are new marvels to see, from primroses in flower as early as January to goldenrod (*Solidago*) and ivy flowering late into autumn. Each habitat has a different mix of species, so a walk at any time of year around the meadows, woodlands and into the Loder Valley Nature Reserve will reward you with new delights.

Hazel and bramble in the woodlands provide ideal cover for rare dormice.

SAVING THE WORLD'S SEEDS

Although the wildflowers of the Weald are flourishing through the sensitive stewardship of Wakehurst, plant species around the world are under threat because of forest clearance, over-exploitation and climate change. As a result, around a quarter of the world's plant species — between 60,000 and 100,000 — face the possibility of extinction.

However, plants are vital to our own survival – they provide the air we breathe, our food, and food for our livestock. They give us clean water, fuel, building materials, fibres for textiles, resins, spices and medicines. They also counteract climate change by removing carbon dioxide – one of the 'greenhouse gases' – from the atmosphere.

The Millennium Seed Bank Partnership (MSBP) works with partners in 80 countries to bank seeds of the world's wild plant species, focusing on the plants that are most at risk and those most useful for the future. The loss of one plant species could deprive future generations of medicine for a disease that is not yet even recognised. It could be the one plant capable of restoring the natural balance of a damaged habitat. It could be a super food able to grow in desert conditions and a valuable source of nutrition for people of the developing world. We can't see the future clearly, so the MSBP and its partner seed banks act as an insurance against the loss of plants.

At the MSBP's striking facilities at Wakehurst, scientists are advancing seed-banking theory and technology. They have already secured over 13% of the world's wild plant species in their vaults and in partner seed banks around the world. By 2020, they aim to have safely stored seed from 25% of the world's bankable plants.

Discover much more about the work of the MSBP in the exhibition inside the Wellcome Trust Millennium Building.

Seeds come in all shapes and sizes and, depending on the environment in which they have evolved, they require particular conditions for long-term storage. Some – like the English oak (*Quercus robur*) – remain a riddle, but most can survive for hundreds of years if they are dried and then stored at -20°C. MSBP scientists develop protocols for the optimum storage conditions for each species, but they also investigate how to break dormancy – to wake the seed up again – so that it germinates and grows into a new plant when the time is right. The seeds of different species need different triggers to bring them into growth – variations of heat, light, water, even smoke and fire.

THE UK NATIVE SEED HUB

On the parterre outside the Millennium Seed Bank (MSB) building, raised beds evoke eight threatened habitats of the British Isles – shingle beach, cliff face, chalk downland, hay meadow, fen and marsh, hill and mountain, heathland and cornfield. These intriguing displays are the shop window of the UK Native Seed Hub project.

To create a nationwide ecological network by expanding and linking surviving habitats requires seeds genetically adapted to the intended site of restoration. Previously it was difficult for conservation groups and landowners to source high quality native-origin seeds and plants through commercial companies. But the MSB holds collections of more than 90% of the UK's native flowering plants and now makes them available to commercial companies for conservation or restoration use through the UK Native Seed Hub. The project started by growing lowland meadow species, as 98% of semi-natural grassland has vanished in England and Wales since the 1930s. Seed production beds were established at Wakehurst to multiply collections of priority species

The rare Dorset heath (*Erica ciliaris*) is one of many threatened native wild plants.

Long beds of pheasant's eye (*Adonis annua*) are grown for their seed here.

such as devil's bit scabious, cuckoo flower, green field-speedwell, pheasant's eye, spiked rampion and harebell, and the project provides seed for direct use in habitat restoration or as starter stocks for commercial seed companies to bulk up for use in landscape-scale conservation projects. Underpinned by Kew's world-class botanical, horticultural, ecological and seed conservation expertise, the project also offers training, technical advice and research to landowners and agencies.

Botanical multi-tasking

Science, conservation and gardening in one job – no problem! Wakehurst is unique in supporting the work of the Millennium Seed Bank Partnership (MSBP). Where else would you find a world-class garden carrying out globally significant science and conservation? To do that, you need a uniquely skilled team.

Working in Wakehurst's nursery is a multi-tasking challenge, with plants raised for the gardens, to add to the tree collections and to support the MSBP. In each member of the team you'll find a person able to switch between the delicate task of pricking-out individual seeds from agar plates to digging up large trees grown for the woodlands.

A CHANGING LANDSCAPE

When the Mansion was built in 1590, Queen Elizabeth I sat on the English throne, the potato had recently been brought back from the New World, the mania for tulips would sweep Europe in 50 years' time and the English Landscape Garden style would not emerge for more than a century.

Imagine the range of gardening this house must have witnessed over the last 425 years! Its residents down the generations were wealthy and connected and would surely have followed fashions set by the aristocracy and gentry, but the development of the gardens would have always been bound up with the extraordinary landscape that surrounds the estate.

In the time of pre-history, floodplains and rivers laid down layers of iron-rich clays and sandstones here. In the middle of the Cretaceous Period, 110 million years ago, these sediments sank beneath the sea. The carcasses of countless sea creatures then deposited a thick layer of chalk on the sea floor. When the African continent crashed into Europe, forming the Alps, the ripples spread northwards and a huge chalk dome rose in what is now south-east England. Over millennia, the chalk top of this dome eroded to reveal a region of sandstone cliffs and steep-sided valleys, known locally as 'ghylls', sandwiched between the chalk hills of the North and South Downs.

This landscape is known as the High Weald. 'Weald' comes from the Old English 'wald' or 'wood' and thanks to its special characteristics it has remained more densely forested than any other part of southern England. The dramatic cliffs, ravines, gullies and ridges mean that the High Weald was unsuitable for large-scale clearance and agriculture, so it was not settled in the same way as the surrounding areas. Instead, the forest was managed to support the iron industry and as a feeding ground or 'pannage' for swine. In late summer, herds of pigs were driven up the ancient lanes to seasonal wood-pasture or 'dens', where they remained until mid autumn, fattened by acorns. Many of these lanes or 'droves' survive as the roads we use today.

Iron ore was plentiful and easily accessible through open cast mines. Iron was extracted by 'smelting', a process that required high-quality charcoal to produce sufficient heat. Even in prehistoric times, people here were coppicing understorey tree species such as hornbeam (*Carpinus*) and hazel (*Corylus*) to produce sustainable harvests of wood to make charcoal. The iron industry remained significant in the Roman era (a Roman road runs northwards under Wakehurst's car park) and saw a resurgence in medieval times. The remains of furnace or hammer ponds used in the industry have been found on the estate near Westwood Lake. Indeed, the landscape of Wakehurst is one of the best-preserved medieval landscapes in Europe.

Gerald Loder

THE MAKING OF THE GARDENS

Two gnarled yew trees on the lawns near the Elizabethan Mansion date back to the earliest incarnations of Wakehurst as an estate (one dates back to 1391 when Richard II was on the throne), and it is thought that several fine specimens of giant redwood (*Sequoiadendron giganteum*) were planted in the 1890s. But it was in the early years of the 20th century when the development of its collections of the world's temperate trees and shrubs really took flight, at the hands of a passionate plantsman, Gerald Loder, later Lord Wakehurst.

Loder, a businessman and member of Parliament, purchased Wakehurst in 1903. He came from a family of enthusiastic gardeners and grew up at High Beeches, another renowned Wealden estate, so he would have been well aware of the horticultural potential of his new home. Loder worked closely with head gardener Alfred Coates, and within five years his plant catalogue detailed some 3,000 species and cultivars at Wakehurst. His 33 years here coincided with an astonishing period of collecting by renowned plant hunters such as George Forrest, Frank Kingdon-Ward and Ernest 'Chinese' Wilson, as well as Wealden man Harold Comber, who made notable collecting trips to Argentina, Chile and Tasmania in the 1920s. Unlike the plant collectors of previous centuries, these men sought plants for horticultural merit rather than economic potential. Loder, who was president of the Royal Horticultural Society from 1929 to 1931, and his circle of influential plant enthusiasts, subscribed to these trips and many gardens that were created during this time, such as Caerhays, Nymans and Bodnant, are still famous today.

Several of the towering redwoods were planted in the 19th century.

After Loder's death, the gardens continued to develop in the hands of businessman Sir Henry Price and his wife, Lady Eve. Noted plant breeders and major exhibitors at Royal Horticultural Society shows, they are remembered in award-winning plants such as *Viburnum* 'Eve Price' and *Pieris* 'Henry Price.' Sir Henry bequeathed Wakehurst to the nation in 1963. Two years later, when Kew took over its management, the transformation into a botanic garden began.

The catalogued plantings created by Loder and Price were the foundations on which Kew began to build. Wakehurst horticulturists including Tony Schilling, Mark Flanagan and Andy Jackson travelled extensively, collecting seed from wild places. Their precious harvest has been tended and nurtured, and great care has been taken with the placement of these wild plants throughout the estate, to fully represent the beauty and variety of the temperate flora of the world, for everyone to enjoy.

Viburnum 'Eve Price' commemorates the work and knowledge of Lady Eve Price.

The Mansion is at the heart of the estate, but it is the collections and native flora – the wild plants from wild places – that make Wakehurst what it is.

Bethlehem Wood

FURTHER INFORMATION

For further information about Wakehurst, its opening times, events, and how to become a Friend please go to **www.kew.org/visit-wakehurst**, email **wakehurst@kew.org** or call **01444 894066**.

FURTHER READING

Bynum, Helen and William (2014). *Remarkable Plants that Shape our World*. Thames & Hudson.

Fry, Carolyn (2012). *The Plant Hunters: The Adventures of the World's Greatest Botanical Explorers*. Andre Deutsch.

Fry, Carolyn, Seddon, Sue and Vines, Gail (2011). *The Last Great Plant Hunt: The Story of Kew's Millennium Seed Bank*. Royal Botanic Gardens, Kew.

Grimshaw, John and Bayton, Ross (2009). *New Trees: Recent Introductions to Cultivation*. Royal Botanic Gardens, Kew.

Hepper, F. Nigel (ed.) (1983). *Wakehurst Place, Yesterday, Today and Tomorrow*. Kew Guild.

Loder, G. W. E. (1907). *Wakehurst Place, Sussex: an account of the Manor and its owners*. Spottiswood & Co.

MacQuitty, Miranda (2011) *Kew at Wakehurst: A children's guide*. Royal Botanic Gardens, Kew.

Price, Katherine (2015). *Kew Guide*. Royal Botanic Gardens, Kew.

Rix, Martyn, (ed.) (2004). *Curtis's Botanical Magazine* Vol. 21, Pt. 1. Wiley Blackwell.

Willis, Kathy and Fry, Carolyn (2014). *Plants from Roots to Riches*. John Murray.

Wood, Carlton and Habgood, Nicolette (2010). *Why People Need Plants*. Royal Botanic Gardens, Kew.

WEB RESOURCES

http://www.highweald.org/
http://highwealdlandscapetrust.org/

PICTURE CREDITS

Images © RBG, Kew: pp 16–17, 18, 25 (panel), 45, 49 (panel), 51 (panel), 52, 59, 60, 61.

Images © Jim Holden: front cover, pp 28, 30–31, 35, 47, 62.

Image © Steven Robinson: p 53.

Images © Paddy Wood: pp 48, 57.

Image © Barney Wilczak: p 46.

Timeline images: Mesolithic arrowhead © John Anderson/Alamy; Iron Age pig © Juniors Bildarchiv GmbH/Alamy; Iron ore © Tom Grundy/Alamy.

ACKNOWLEDGMENTS

We would like to thank Chris Clennett, gardens manager at Wakehurst, for researching and writing the majority of this guide and for providing nearly all of the images within its pages. Thank you also to Katherine Price for her additional writing and advice, and to Jim Holden, Paddy Wood, Steven Robinson, Barney Wilczak and the photographers at Kew for use of their images.

The following have provided valuable help, advice and support: Andy Jackson, Catharine Pusey, Jo Wenham, Gina Fullerlove, Christina Harrison, Georgina Smith and Lydia White. Finally, thank you to Christine Beard for typesetting and layout design.

First published in 2015 by
Royal Botanic Gardens, Kew
Richmond, Surrey, TW9 3AB, UK
www.kew.org

ISBN 978-1-84246-607-0

Distributed on behalf of the Royal Botanic Gardens, Kew
in North America by the University of Chicago Press,
1427 East 60th Street, Chicago, IL 60637, USA.

British Library Cataloguing in Publication Data
A catalogue record for this book is available from the British Library.

Text: Chris Clennett and Katherine Price
Typesetting and page layout: Christine Beard
Project management: Georgina Smith
Copy editing and proof reading: Christina Harrison
Picture research: Chris Clennett and Christina Harrison

For information or to purchase all Kew titles please visit shop.kew.org.
or email publishing@kew.org

Kew's mission is to inspire and deliver science-based plant conservation
worldwide, enhancing the quality of life. Kew receives half of its running
costs from Government through the Department for Environment, Food
and Rural Affairs (Defra). All other funding needed to support Kew's vital
work comes from members, foundations, donors and commercial activities
including book sales.

Printed in Great Britain by Short Run Press